Simple Methods
of
Making Increase

by Wally Shaw

NORTHERNBEEBOOKS.CO.UK

Simple Methods of Making Increase
© Wally Shaw

ISBN 978-1-908904-75-1

Published by Northern Bee Books March 2015
with agreement of Cymdeithas Gwenynwyr Cymru -
The Welsh Beekeepers Association

Scout Bottom Farm
Mytholmroyd
Hebden Bridge HX7 5JS (UK)
www.northernbeebooks.co.uk

Design and Artwork gustav m

Printed by Lightning Source UK

contents

1. Introduction

The aim of this booklet is to give detailed coverage of this important subject and in a form of a practical guide for use by both individual beekeepers and associations who want to become **self-sufficient** for the provision of new or replacement colonies and queens. The methods described are not designed for large scale queen rearing but should be more than adequate to meet the needs of the hobby beekeepers who, let us not forget, manage about 85% of the colonies in Britain.

Rather than provide cut-and-dried recipes, the aim (of this booklet) is to explain the principles (the bee biology) on which making increase is based. If the principles are understood the beekeepers can tailor the details to match a wide range of situations; the resources (the colonies that are available), the amount of increase required, the time of year and the equipment available. Also to be taken into account is the degree of pre-emptive swarm control which comes as a useful bonus when making increase. Experience has shown that using methods that are more closely akin to the natural behaviour of a honey bee colony (what it has evolved to do) has many practical advantages and results in a high level of success. Accordingly some departures from conventional practice are described. At the end of the booklet a number of practical examples are given as general guidance and to illustrate the flexibility that is possible.

2. The Locally Adapted Bee

One of the reasons why the honey bee is so successful over such a wide range of climatic regimes (from both the wet and dry tropics, through sub-tropical to cold temperate climates) is that natural selection has created races (or strains) that are adapted to the conditions where they live. These are what are now termed 'locally adapted bees' and it is becoming widely accepted that, for all sorts of reasons, these are preferable to bees from outside sources. In this context, 'outside sources' refers particularly to bees imported from other countries but also includes those from other parts of Britain where different climatic conditions prevail. In the extreme oceanic climate of Wales, a strain of bee containing genes predominately derived from the Northern Dark Bee (syn. the Welsh Black bee – *Apis mellifera mellifera*) is likely to be the best adapted bee for most areas.

Apis mellifera remains a single species in the sense that all the sub-species

Wally Shaw

are genetically compatible and can inter-breed. As a result of past introductions of non-native races there is usually some degree of 'mongralisation' (genetic introgression) in the locally adapted bee. The fact that this condition tends to persist over a number of generations implies that the genes involved confer some advantages (they are adaptive). Our aim should be to achieve a degree of genetic stability in our local bee; in other words to permit natural selection to 'weed-out' non-adaptive genes. Each time we introduce bees from outside sources it is like taking a backward step in time. The dispersion of drones and loss of swarms from non-local colonies creates 'genetic ripples' that persist until they are again resolved by natural selection.

3. Reasons for Learning How to Make Increase
Making increase is a basic skill that all beekeepers should possess but at the present time we fall well short of this target. Many beekeepers rely solely on swarms to provide new colonies but this is rather hit-or-miss and can not be done to order. There are many reasons, both for the individual beekeeper and for beekeeping as a whole, why this situation should be rectified.

3a. For the individual beekeeper:-
- To increase their number of colonies on demand (and at little cost).
- To replace winter losses.
- To provide colonies for other people - particularly starter colonies for new beekeepers.
- To improve the characteristics of their bees – health and productivity being the most important.
- For interest and satisfaction.
- As a vehicle for increasing their beekeeping skills.

3b. For the good of beekeeping and the welfare of the honey bee:-
This is mainly about reducing the number of colonies and queens that are imported (to Britain) and keeping locally adapted bees (see above).
- Most of our current problems in beekeeping are the result of globalisation – moving bees around the world.
- There is still potential to create more problems, eg, the introduction of exotic pests such as small hive beetle and Tropilaelaps.
- To limit the spread of new strains of endemic diseases with different levels of pathogenicity. The veterinary certificates that are compulsory with **legal** importations are only a limited safeguard.

Simple Methods of Making Increase

Because of free-trade policy, becoming **self-sufficient** for the production of new colonies and queens (thus reducing or eliminating the demand) is the only realistic way of achieving a reduction in imports. There is a regular demand for early season queens because many beekeepers are unwilling to accept a delay of about 4 weeks in order to raise new colonies for themselves. Unfortunately the British climate precludes the production of home-grown queens early in the season so this will require a change in beekeeping culture. Beekeepers should learn to do more re-queening in the autumn and have contingency plans to meet what are often unpredictable circumstances, eg. high winter losses.

4. Scales of Increase Covered

As already noted the methods described in this booklet are not suitable for making increase or queen rearing on a large scale but cover:-

- Simple 1 colony increased to 2 (or possibly 3).
- Making 3-5 colonies (nucs.) from 1 colony at a given time.
- Making increase can usually be repeated 4-5 weeks later when the 'breeder-colony' has re-built, giving a maximum of 8-10 new colonies from an individual colony in a year.

Small scale increase (using a number of breeder colonies) has the important advantage of maintaining genetic diversity whereas larger scale schemes (using grafting) can have the opposite effect – especially if the resultant colonies are not disbursed over a wider area. The methods described below all depend on inducing a colony (or part of a colony) to produce **emergency queen cells**. This is one of three circumstances in which colonies produce queen cells and it is important to understand the different causes and outcomes of what are **distinct** behavioural programmes:-

1) **Swarming** – the colony wants to reproduce and the outcome is **a swarm (or swarms)**.
2) **Supersedure** – the colony wants to replace an old or failing queen, there is **NO intention to swarm** and the outcome is more or less seamless replacement of the existing queen.
3) **Emergency re-queening** – the colony has lost its queen and, providing she has laid eggs within the last 5 days, the colony is able to make emergency queen cells. The colony will automatically select a **(single)** replacement from however many queen cells that are produced and will make NO attempt to swarm.

Wally Shaw

5. Prejudices over Emergency Queens

It is a widely held view that emergency queens are inferior to swarm queens and this is reinforced by many (but not all) beekeeping books and articles. Grafting of young larvae into cups so that they can be raised by first a starter colony and then by a finisher colony is seen as a more reliable (the standard) method of producing top quality queens. However, it should be clearly understood that this technique does NOT produce true swarm queens because nowhere in the process is the swarming impulse involved.

The tarnished reputation of emergency queens is probably based on last-ditch attempts (by the beekeeper) to re-queen a colony that has been without a queen for 6-8 weeks (or more) by introducing a frame with eggs and/or young larvae sourced from another colony. By now the colony is composed of elderly bees with limited ability to produce high quality brood-food that is necessary to raise a fully developed queen. Another reason why emergency queen cells are regarded as inferior to swarm cells is that they are usually set into the face of the comb and look less impressive. Despite their outward appearance, emergency queens are raised in vertical cells that have the same internal volume as swarm cells (see Figure 1).

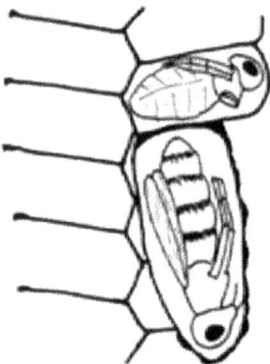

Figure 1 - Comb modification used to construct an emergency queen cell

The final proof of the reliability of emergency queens comes from scientific studies which show that they are anatomically identical to swarm queens; they have the same number of ovarioles (the egg producing units) and produce colonies with the same level of performance as those raised from swarm cells. All that is required to produce a fully developed emergency queens is that the colony in which they are raised has adequate resources of bees (and particularly nurse bees of all ages), brood and food. The whole future of a colony depends on producing a good queen and that is exactly what it will do given half a chance.

6. Another Common Misconception

Many beekeepers think that a colony with multiple queen cells will swarm – or is likely to do so. Beekeeping books send out a mixed message on this matter and it is often recommended that beekeepers should thin queen cells (usually to a single cell) in circumstances when this is totally unnecessary. Thinning queen cells is an activity to be avoided wherever possible because it requires the beekeeper to make a critical choice on which the survival of the colony depends. The same applies to choosing single queen cells (deciding which is 'best') to furnish a nuc. The beekeeper can only judge on the external appearance of the queen cell and knows nothing about its contents. From their behaviour (as seen in observation hives) the bees show an intense interest in maturing queen cells and it is the workers that do at least the initial thinning of queen cells. The choice of a new queen is clearly a matter where it should be assumed that the **'bees know best'**.

Only a colony that has produced queen cells under the swarming impulse will issue a swarm (or swarms). The only circumstance in which it is (ever) necessary to thin queen cells is with a colony that has already issued a prime swarm when this measure is essential to prevent cast swarming. By contrast, a colony producing emergency queens or a queen-less colony given queen cells (eg. when making a nuc) has no intention of swarming and will automatically select a new queen from the cells available - however many their number.

7. Making Emergency Queen Cells
As already noted, the methods of making increase described in this booklet are based on the use of emergency queen cells. All that is required to induce a colony to make emergency queen cells is to remove the queen. But, having done this, what do you do with her? The existing queen is the beekeeper's most valuable asset and the only way to combine the dual aims of making increase and obtaining a honey crop is to split colonies in a balanced way. Splitting colonies is also by far **the most effective** method of preventing colonies setting up to swarm. It follows from this that making increase and pre-emptive swarm control are merely the two sides of the same coin and should be closely integrated activities in the beekeeper's management programme.

8. Why not wait until a colony sets up to swarm?
You can do this but it has a number of disadvantages:-
* In order to prevent swarming the beekeeper must create some sort of artificial swarm. Because this entails the segregation of flying and non-flying bees, the split will be less well-balanced than if it had been done pre-emptively.

Wally Shaw

- Initially the resultant colonies will have an abnormal worker age-class distributions and it will take several weeks for this to be normalised. For this reason an artificial swarm has less potential for honey production than a controlled split.
- Nucs made out of a colony that is set up to swarm **MUST** remain in the same apiary long enough for the flying bees to return to their original hive position. If a nuc is immediately moved to an apiary at a distance, from which the flying bees can not return home, the swarming impulse persists and it is likely to swarm with the first virgin queen to emerge.
- Also when populating such nucs it is difficult to judge how many bees is enough because the number that will be lost is unknown (and very variable) – it is necessary to err on the side of caution.
- In addition, you **may** be inadvertently selecting for swarmy bees.

9. When to Split Colonies

Deciding when to split a colony requires a number of important factors to take into account:-

- The size and development of the colony - its resources in terms of bees, brood and stores. Ideally a colony should be close to a condition in which it might be expected to set up to swarm.
- The time of year – plentiful drones available and the expectation of suitable weather for queen mating (late-April through to the end of June is the optimum period in most areas).
- Timing in relation to nectar flows – at the end of the spring flow (or after oil-seed rape) minimises the impact (if any!) on honey production.

10. What Must the Colony Have to Make an Effective Split?

Apart from plenty of bees, brood and stores (see above) the queen-less part must:-

- Have plenty of eggs and young larvae present.
- These should preferably be in new or young combs - old black combs (with several layers of pupal skins lining the cells) are more difficult for the bees to structurally modify to make emergency cells.
- After the flying bees have departed (they go back to join the old queen at the original hive location) there must be sufficient bees remaining to cover the brood. Providing the brood is kept warm emerging bees will quickly supplement their number.

11. Which Colonies to Split

All sorts of criteria (often using elabourate scoring systems) have been devised to select the colonies from which to breed (make increase). There is no reason why you should not use one of these if you so desire. However, this is a short list of the more important selection criteria:-

- Health and vigour – resistance to diseases such as chalk brood and Nosema.
- Productivity – hard-working and prolific enough to make a good sized colony.
- Responsive to local climate – do not over-produce brood under adverse conditions
- Reasonable temperament – slightly 'prickly' colonies are *sometimes* very productive and you may wish to compromise on this criterion.
- Not easily triggered to swarm – some of the least swarmy bees produce small colonies, so again compromise may be necessary.

12. How to Balance the Split

There is no fixed recipe for making the splits because it depends on the colony that is going to be split and what the beekeeper wants to achieve:-

- The configuration of the hive (the brood area) – single deep, brood and a half, double deep or extra deep.
- How many frames of brood are available – which is partly dependent on hive configuration.
- The aims of the split – whether it is an increase of 1 to 2 or to make as many nucs as possible.
- The importance of the swarm control element.

Two box systems (brood and a half or double brood) are more flexible than single box systems (deep or extra deep). Brood and a half probably offers most options, with up to 24 frames available in two different sizes. Even if you normally use a single brood box it could be an advantage to convert (temporarily) to a two box system for making increase (see Example 3).

The only (essential) rule about splitting hives is that **both parts of the split must be viable.** An example of a safe minimum split would be to transfer 3 frames of brood and 2 of food into a nuc box and leave them to make a new queen. This is a well-established practice for beekeepers whose main crop comes from heather, the nuc and the parent colony being re-united just before

Wally Shaw

going to the heather. However, **a more generous split is preferable to ensure the production of high quality emergency queen cells.**

The parent colony (that remains on the old site and contains the queen) will automatically acquire most of the flying bees but will also need to have enough nurse bees (and a continuity of supply in the form of brood) to enable the queen to quickly lay the empty combs with which she has been provided. Some or all of the supers usually remain with the parent colony and this provides an additional source of younger bees that are still capable of adapting to nursing duties.

The daughter colony (on a new site and queen-less) will initially be short of flying bees but will rapidly promote some of the older house bees to this role. In 9-10 days time there will no longer be any unsealed brood to feed and there will be sealed emergency queen cells only a few days away from maturity. As long as there are sufficient bees to cover the brood (and some to spare) the colony will continue to forage and at least feed itself. There are usually frames of food that can be given to this part of the split or, failing this, it can be given a honey super (or part of one). Only if there is a run of poor weather is supplementary feeding usually required but it is advisable to err on the side of safety.

It is often said that splitting a colony will cost you a super of honey and this is a disincentive for many beekeepers. However, experience of splitting hives shows that it has less effect on honey production than you might think. A well-balanced split gives the queen a lot of empty frames to lay and, providing she has the support of sufficient nurse bees, she will go into 'overdrive' and produce more brood than if the colony had not been split. In the case of a 1 to 2 split, if it is made at the right time (after the spring flow), the parent colony (with the old queen) and the new colony (with a new queen) will both have a sufficient number of bees to make a substantial honey crop from the main flow. The combined yield may be greater than if the original colony had remained intact and certainly greater than if it attempted to swarm (and had been artificially swarmed) or, worse still, actually succeeded in swarming.

13. A More 'Natural' Approach to Making Nucs.
When a colony is re-queening after swarming or the beekeeper has made a 1 to 2 split, there is a high probability (90%+) of producing a new queen with a normal reproductive life ahead of her. This high level of success contrasts with

the outcome if mini-nucs are used, where 2 out of 3 (66%) success rate is about the best that can be expected - and it can often be substantially lower - so why the difference?

It is partly a matter of scale, because (in a mini-nuc) as few as 200-300 bees supporting a virgin queen would not occur naturally – even the smallest cast swarms are much larger than this. Other departures from what is natural that beekeepers often inflict on bees when making increase (see below) may also contribute to poor results.

If the aim is to make a 1 into 2 increase the task is completed as soon as the split has been made. The parent colony (the one that retains the old queen) and the daughter colony (the one that is making emergency queen cells) can remain in the same apiary, either on separate stands or with the daughter colony on top of the parent colony using a split board. All the beekeeper needs to do is sit back and wait and, after 4-5 weeks, check that a new queen is laying in the daughter colony. The parent colony should have taken a sufficient 'hit' (in terms of bees and brood) and is unlikely to swarm.

Setting-up nucs using the emergency queen cells that have been produced by the split is a little more complicated and requires some attention to detail if good results are to be obtained. By observation, trial and error over several years we have devised a system for producing nucs which has a high success rate with about 90% raising good, long-term queens (surviving at least to end of the following season). The key practices involved (please note these are not to be found in any beekeeping books) can be summarised as follows:-

1. The nucs should be populated with bees from the **same colony** that made the queen cells.
2. Each nuc should be given **at least 2 queen cells** and preferably more (there is no limit), the cells being transferred *in situ* on the frames on which they were built – **NO** selection of queen cells is required.
3. Frames with queen cells on them should be harvested **as soon as possible after sealing** (about day 9).
4. **The entrances to a nuc boxes should be blocked** whilst being populated with bees and brood (to retain as many bees as possible) and should then be removed to a mating apiary that is 2-3 miles distant.

Wally Shaw

We think this works because it is more 'natural', ie. it closely resembles what happens in nature, and the explanation is as follows:-

Guidelines 1 and 2 are logically related and are aimed at ensuring that the developing queens are tended by worker bees to whom they are genetically related (1). Bees are very much aware of the genetic relationship with their nest mates (whether they are full or half sisters – depending on the drone that fathered them). These relationships are thought to be an integral part of colony organisation, albeit they are not fully understood at the present time. This contrasts with the traditional method of making nucs, where a single queen cell is usually given to a bunch of non-related bees. Provisioning a nuc with multiple queen cells has the advantage that the bees can do 'what comes naturally' and exercise choice (2). We (mere humans) do not understand what criteria they are using but **they** (the bees) clearly do and the best strategy is not to interfere.

Guidelines 3 and 4 are also logically related and are aimed at increasing the chances of the virgin queen mating successfully. Making up the nucs as soon as the queen cells have been sealed (3) is done to maximise the time (number of days) before a new queen will have emerged and be ready to take mating flights. Blocking of entrance prevents the loss of flying bees and removal to another (distant) apiary (4) ensures that they stay with the nuc and can not fly back home. Recent studies show that virgin queens do not go out alone on their mating flights but are accompanied by a number of mature worker bees who guide her to the drone assembly areas and ensure her safe return. In order to do this task well a nuc must have plenty of forager bees who have had sufficient time to learn their new territory. If nucs are made up on day 9 and immediately moved to the mating apiary there should be a minimum of 7 days before a new queen is ready to mate (and probably longer) and the flying bees are called on to perform this important function. Lack of bees may be one of the reasons mini-nucs have lower level of mating success than larger colonies. An additional disadvantage may be that poorly supported queens are forced to mate nearer to home resulting in reduced access to genetically diverse drones.

14. Nuc Boxes

Unless it is intended to try and over-winter colonies in nucs, the boxes used to make increase can be quite simple (but must incorporate correct bee-space) as they will only be in use during the warmer months of the year. We do not recommend the use of mini-nucs (for the reasons given above) but nuc boxes that will accommodate shallow frames are a useful addition to the available equipment. Their advantages are as follows:-

a) They are economical with their use of bees and brood.
b)   But large enough to hold a properly balanced colony to support the queen during the mating process and care for her until such time as her own progeny take over the role.
c) They enable making increase to be more flexible and help minimise the loss of honey yield.

Shallow nucs can initially be hived-up in a shallow box with a deep box being added at a suitable point in their development. It works as well as deep nucs; they just take a bit longer to build-up.

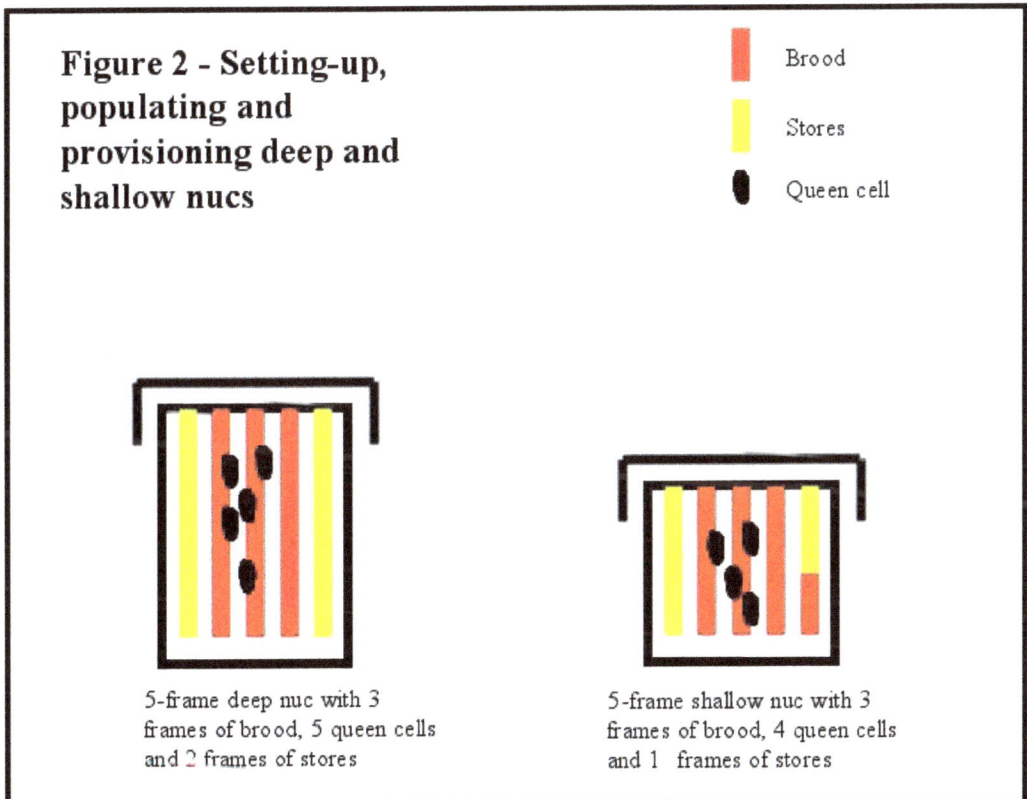

Figure 2 - Setting-up, populating and provisioning deep and shallow nucs

Brood
Stores
Queen cell

5-frame deep nuc with 3 frames of brood, 5 queen cells and 2 frames of stores

5-frame shallow nuc with 3 frames of brood, 4 queen cells and 1 frames of stores

Wally Shaw

15. Split Boards

This is simply an intermediate floor that can be used to divide a colony into two independent parts. Split boards have been used in beekeeping since the late 19th century and there are many different designs. The basic feature is that the board isolates the 2 parts of a colony (so that bees can not move from one to the other) and this means that it has to incorporate a separate entrance on its upper side. The usual rules of bee-space must be observed both below and above the board but it is more satisfactory for the bees (freedom of movement) if there is extra space on top of the board and 9mm is recommended for a bottom bee-space hive or 15mm for a top bee-space hive. The board should also have mesh-covered hole (80 x 80mm is about right) in the middle. This hole allows warmth to come up from the colony below and also maintains a common hive smell which facilitates re-uniting if required. The Snelgrove board (with its 8 controllable entrances) is probably the ultimate, all-purpose split board but for the purpose of making increase one entrance and a ventilation hole is all that is required. Using a split board has the following advantages:-

- The split does not need a new hive stand.
- It economises on equipment – no separate floor, cover board or roof required.
- The part of the split on board is warmer and can contain less brood and bees.
- It facilitates re-combination if required.

The downside of using a split board is that the bottom part of the hive can not be inspected without removing the top part. In practice, the pre-emptive swarm control afforded by splitting the colony means that the bottom part requires little attention for 4-6 weeks, providing it has enough space to store honey – extra supers should be added at the time of the split (see Example 3b).

16. Finding the Queen

Virtually all hive-splitting operations start with finding the queen so before proceeding to look at examples of splits it is useful to cover this topic. Having to find the queen is regarded as a serious obstacle by many beekeepers but it is really only a matter of practice. Going about the task in a calm and orderly manner helps, and here follow some practical suggestions.

It helps if the queen is already marked and this is best done early in the season (usually at the time of the first inspection) when the colony is still of modest size and the queen is likely to be hard at work laying eggs on a frame. If the hive has more than one brood box, the process of finding the queen is easier using good logistics. The brood boxes should be separated and searched independently and it is a good practice (and saves a lot of hassle) to use an 'Inspection board'. This is simply a cover-board (or any board with a bee space on at least one side) which has had any holes blocked with mesh. This board is placed on the upturned roof and the top brood box is removed and placed on it before commencing to look for the queen. This should be done with as little disturbance as possible and with the cover-board still in place. An 'Inspection board' has three purposes:-

1) To minimise disturbance to the bees.
2) To avoid the queen falling out and possibly being lost.
3) To prevent the bees hanging down in festoons beneath the frames – which is what happens when a box is inspected over a void.

Wally Shaw

Plan A – this is finding the queen by a thorough frame inspection and is just a matter of practice (brain-training rather than eye-sight). With the brood boxes separated, search the most likely box first (your hunch) using as little smoke as possible. If she is not found proceed to search the other box. If you do not find her at the first pass repeat the whole process. By now there may be clues as to which box contains the queen because the bees in the box where she isn't will start to run around looking for her. Occasionally a queen proves impossible to find and this is probably because she is no longer on a frame but has taken up refuge on the hive wall or floor. Now there are two options; firstly, put the hive back together and try again later (there is not the same urgency with a spit as there is with an artificial swarm) or secondly resort to Plan B.

Plan B – this is a variant of the shook swarm; ensuring that ALL the bees currently in the brood area (which must include the queen) are in the box where the queen needs to be when the split is complete ie. where she would have been placed if Plan A had worked. The box (or boxes) that she needs to be in should be placed on a floor (normally this is on the original hive stand). All the bees on the frames that are destined to go to the other (queen-less) part of the split are then shaken into this box (or boxes), not forgetting to carefully brush the bees off the hive wall. Good logistics are essential and it is useful to have a spare empty box in which to place the shaken frames – most of which will contain brood needing sympathetic handling. When all the bees from the brood area (including the queen) are in the box (or boxes) that are to form the queen-right part of the split, it is just a matter of re-populating the brood on the shaken frames with nurse bees. A queen excluder is placed over the box (or boxes) in which the queen now resides and the box containing the shaken frames is placed on top (the supers are then added to keep the bees in them happy). Smelling the brood above, nurse bees will quickly move up to cover the shaken frames. One hour is usually enough time for this process to be completed, so do something else in the apiary or go and have a cup of tea. On return, the queen is where you want her, the shaken brood frames have a full complement of nurse bees and it is simply a matter of completing the split.

Wally Shaw

17. Details and Discussion of Examples

Example 1 – the starting point is a hive on a brood and a half (shallow brood on top of the deep) with one honey super above a queen excluder - a fairly typical hive in early to mid-May. There are 10 frames of brood in the deep box and 8 in the shallow. The first step is to find the queen (see above) and when found she can either be placed in a cage (a plunger type queen marking cage is useful here) or transferred directly into the shallow brood which, in this example, is destined to go with the queen-right part of the split. The deep brood frames are re-arranged as shown in the diagram, with 2 placed in the new deep box (green) at the bottom of the queen-right part and the remainder (8 frames) in the deep box that forms the queen-less part of the split. The queen-less part is moved to a new hive stand nearby. The aim of this example is to create a 1 = 2 split but alternatively, when the queen cells that are produced in the queen-less part of the split are sealed, the frames could be further sub-divided to make up nucs. Depending on the number of frames with queen cells and the number of bees in this box 2-4 nucs could be produced – following nuc making guidelines 1-4 above and shown diagrammatically in Figure 2.

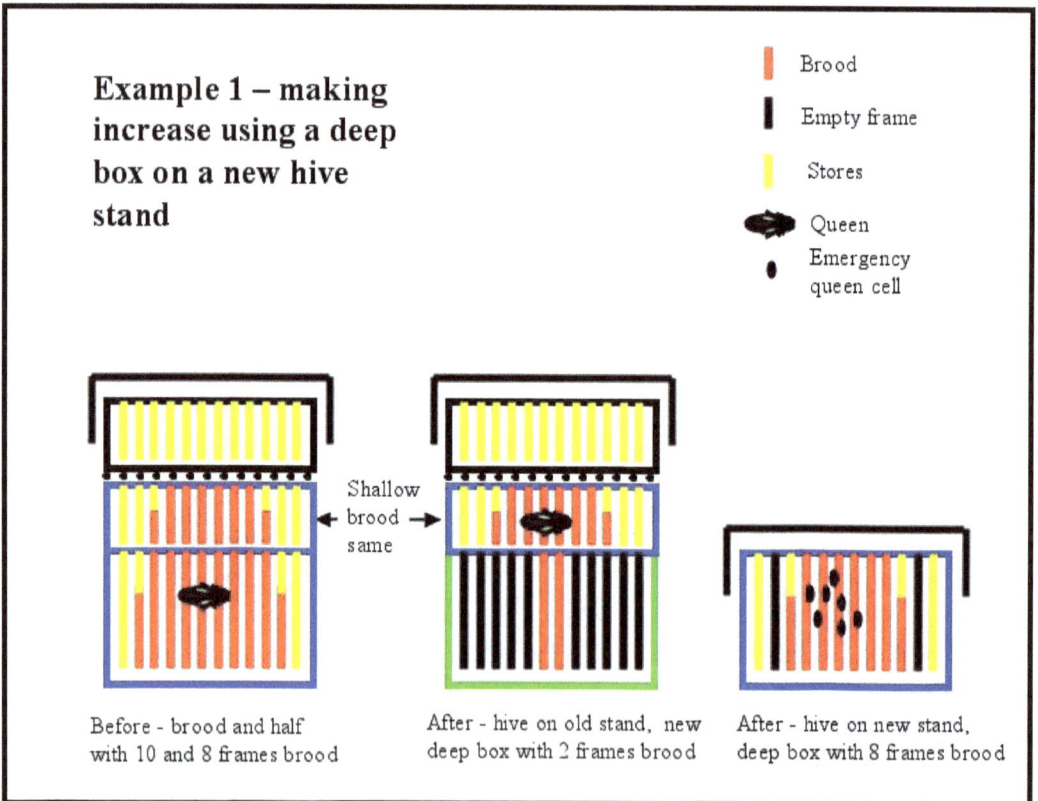

Example 1 – making increase using a deep box on a new hive stand

Brood
Empty frame
Stores
Queen
Emergency queen cell

Shallow brood same

Before - brood and half with 10 and 8 frames brood

After - hive on old stand, new deep box with 2 frames brood

After - hive on new stand, deep box with 8 frames brood

Simple Methods of Making Increase

Example 2 – is similar to Example 1 but this time the queen-less part has been placed on a split board at the top of the hive. The advantages of using a split board are that it requires less equipment, if the split fails to make a new laying queen it can more easily be re-united with the original colony below. Although not shown, the warmth on split-board allows the allocation of brood to be less radical and more can be retained in the queen-right part of the split. Using this method a 50/50 split, with 5 deep brood frames allocated to each part would be perfectly viable. This solution would enable the bottom part of the colony to recover more quickly and yield more honey. The downside is that a reduced number of queen cells and bees in the queen-less part would have less potential for producing nucs. You can't have it both ways and it depends on the beekeeper's priorities – making increase or producing honey?

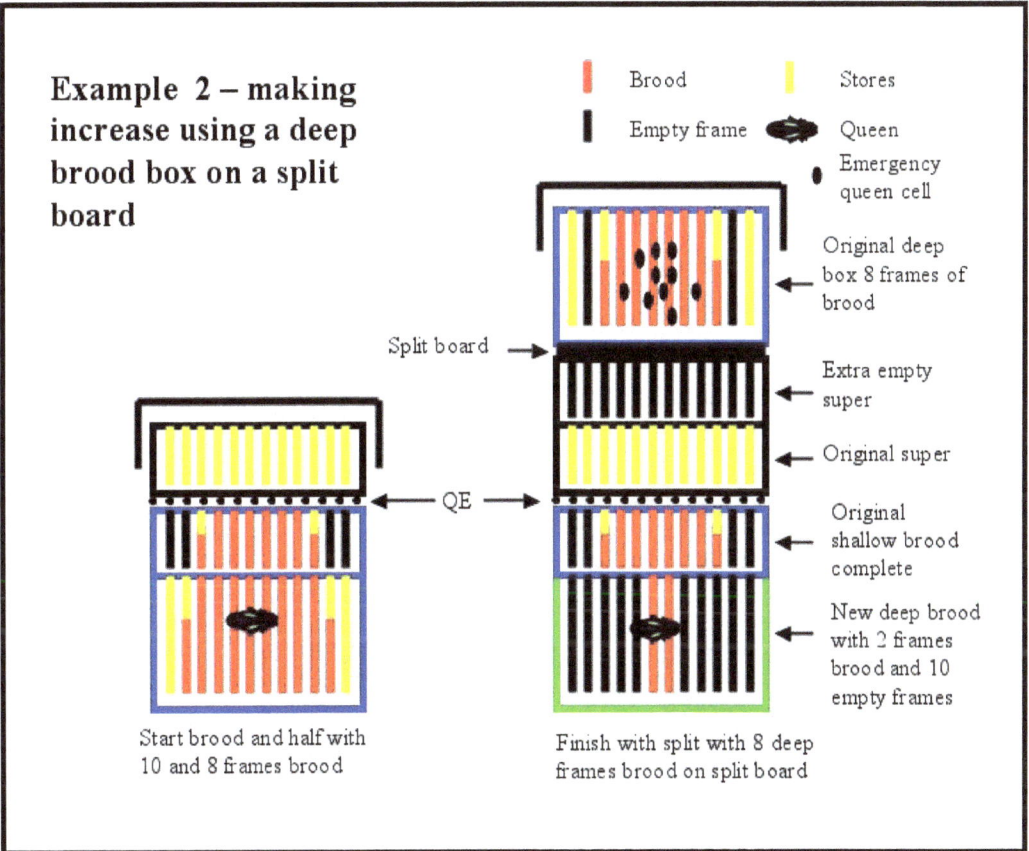

Example 2 – making increase using a deep brood box on a split board

Brood   Stores
Empty frame   Queen
Emergency queen cell

Split board →

Original deep box 8 frames of brood

Extra empty super

Original super

QE

Original shallow brood complete

New deep brood with 2 frames brood and 10 empty frames

Start brood and half with 10 and 8 frames brood

Finish with split with 8 deep frames brood on split board

Wally Shaw

Example 3a and b – shows what is, in effect, a double split with a recovery period of 4-5 weeks between the two stages. This is a more ambitious example of making increase in which the first split could yield 3-4 deep nucs and the second split 3-4 shallow nucs. In 3a the whole of the deep brood with 10 frames of brood is put to the top of the hive on a split board. The shallow brood with its 8 frames of brood remains at the bottom of the hive and fresh laying space for the queen is provided by the addition of two shallow brood boxes containing empty drawn frames (it can be done with just one shallow box). If the brood in the shallow box is a bit sparse the balance of the split can be improved by shaking the bees from a couple of deep frames into the bottom of the hive – this will assist the queen to lay the new space available to her more quickly. When the queen cells produced in the deep box are sealed this can be completely split into nucs and any left-over bees can be allowed to join the colony below.

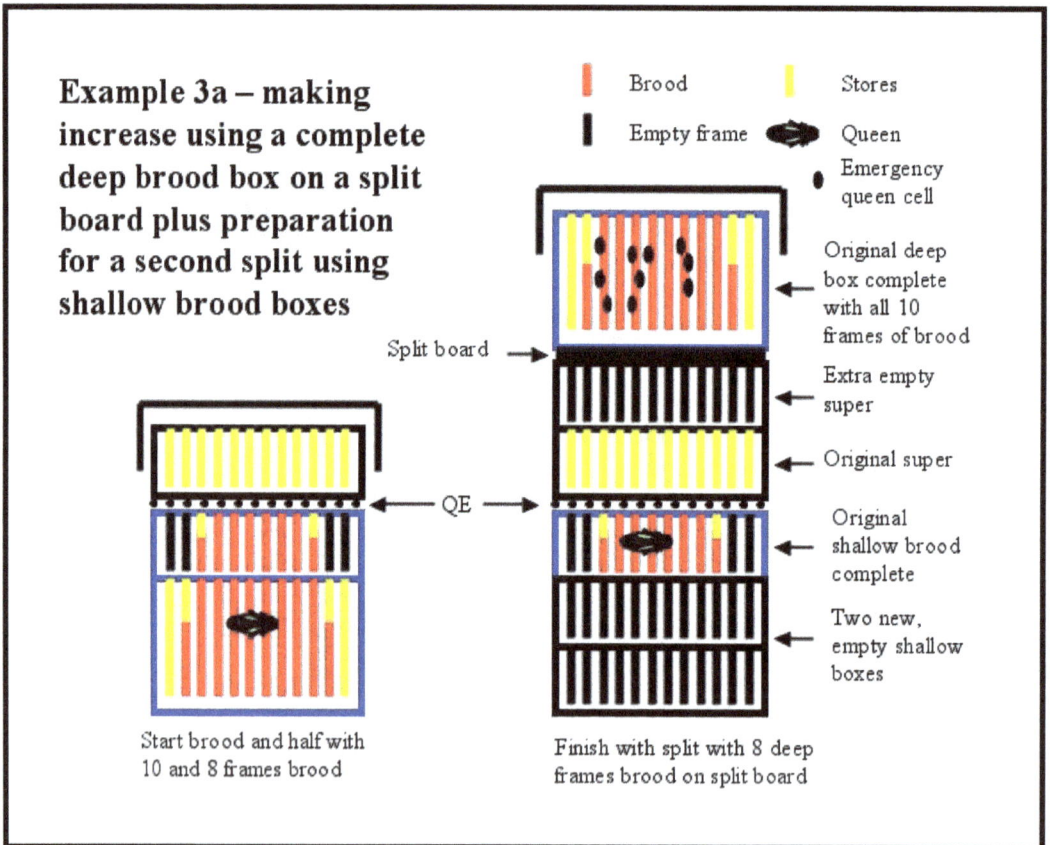

Example 3a – making increase using a complete deep brood box on a split board plus preparation for a second split using shallow brood boxes

Brood   Stores
Empty frame   Queen
          Emergency queen cell

Original deep box complete with all 10 frames of brood

Split board →

Extra empty super

Original super

QE

Original shallow brood complete

Two new, empty shallow boxes

Start brood and half with 10 and 8 frames brood

Finish with split with 8 deep frames brood on split board

In 3b, when the shallow boxes have been comprehensively laid, the hive is split again with all the shallow boxes going to the top of the hive on a split board. The queen remains in a new deep box at the bottom where she again starts to re-build that part of the hive. A 'softer' alternative would be to re-arrange the brood in the three shallow boxes placing the best frames (for making queen cells) in two of the boxes and the third box could then be returned to the bottom part of the hive to help the queen there to re-build more quickly. When the shallow boxes contain frames with sealed queen cells they are in turn divided between as many shallow nucs as the resources will allow (3-5 nucs is usually possible).

During the 4-5 week development period it may be necessary to add further supers beneath the split board.

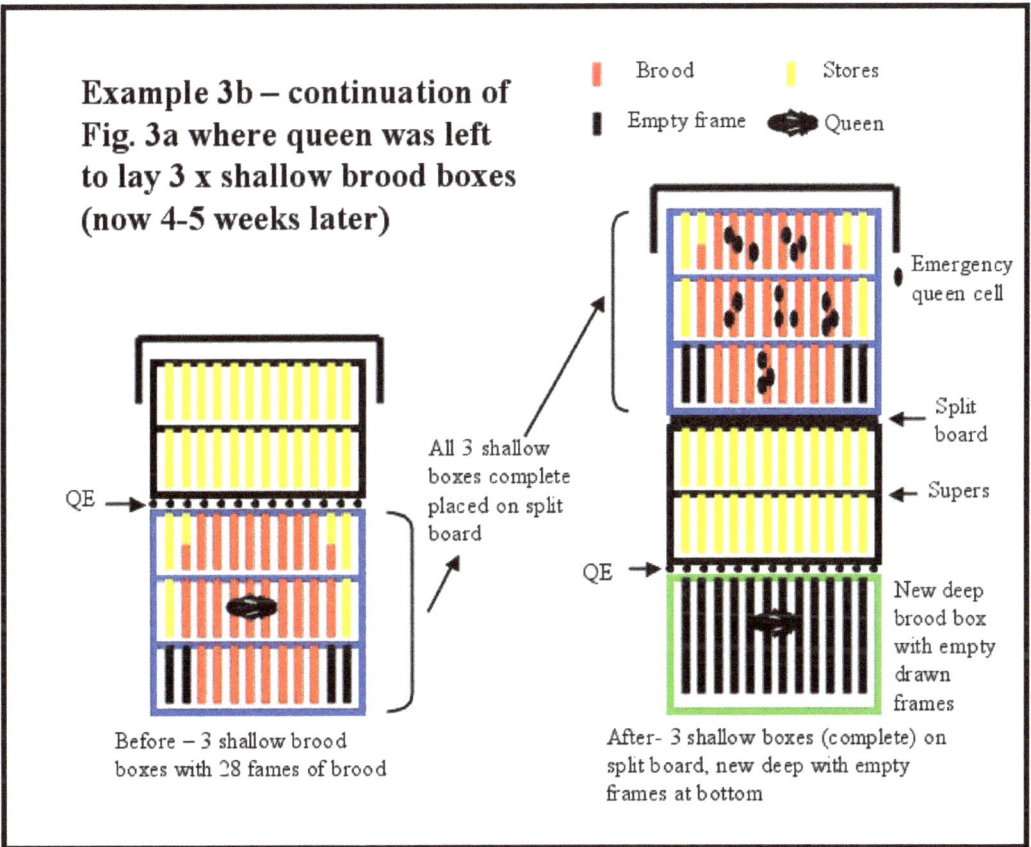

Example 3b – continuation of Fig. 3a where queen was left to lay 3 x shallow brood boxes (now 4-5 weeks later)

Brood   Stores
Empty frame   Queen

All 3 shallow boxes complete placed on split board

Emergency queen cell

Split board

Supers

New deep brood box with empty drawn frames

Before – 3 shallow brood boxes with 28 fames of brood

After- 3 shallow boxes (complete) on split board, new deep with empty frames at bottom

Wally Shaw

18. Drawn Comb or Foundation?

The examples above all specify that the number of frames in boxes should be made up using drawn comb. So what do you do if no drawn combs are available? With simple splits like 1 colony into 2 the use of foundation instead of drawn comb is acceptable (because there will be plenty of bees to do the work of drawing) but, unless there is a sustained nectar flow in progress, it is advisable to feed in order to speed the process.

If drawn combs are in short supply (and this is a common problem for beginners and those who want to make a substantial increase in their number of colonies) the best option is to get them drawn before making the split. Drawing foundation in advance of the split means there are more bees to do the work and it is easier to generate the increased temperature (about 42°C) at which wax can be manipulated. It also means that there will be more frames of brood when the time comes to make the split and it should prevent a premature attempt to swarm. A simple method by which new combs can be drawn early in the season is described next.

Getting Combs Drawn Prior to Making Increase

Principles. The accompanying diagrams show a second deep brood box with

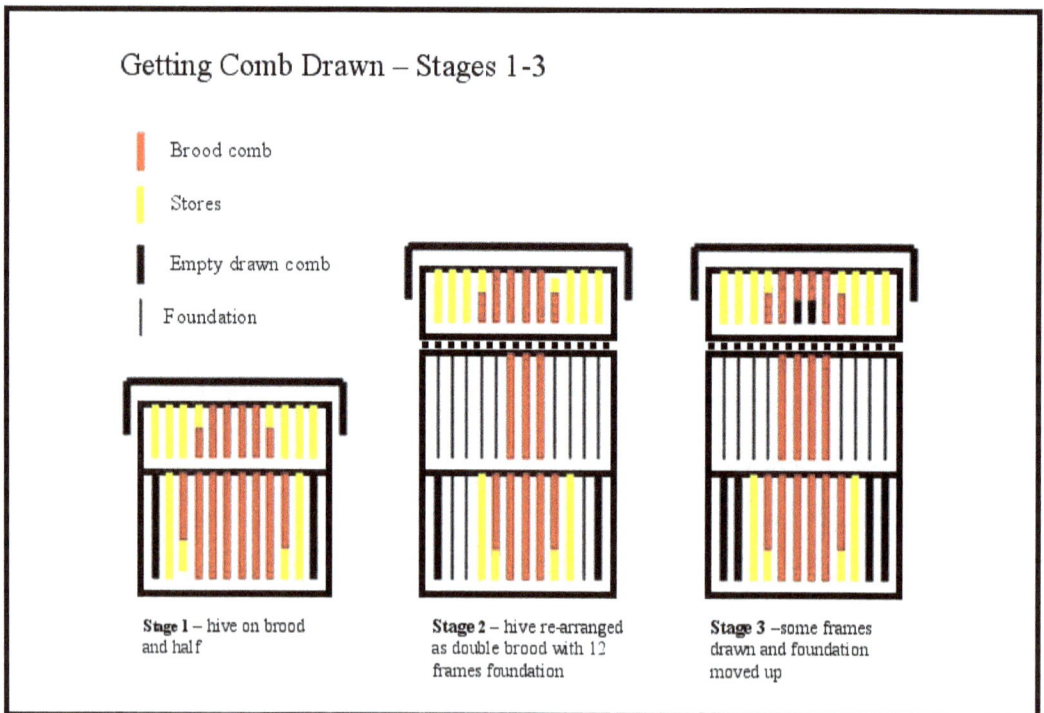

Getting Comb Drawn – Stages 1-3

Brood comb

Stores

Empty drawn comb

Foundation

Stage 1 – hive on brood and half

Stage 2 – hive re-arranged as double brood with 12 frames foundation

Stage 3 – some frames drawn and foundation moved up

Simple Methods of Making Increase

12 frames of foundation being added to a hive. Stages 1-5 illustrate how the existing brood, stores and any empty comb can be progressively re-arranged to get the foundation drawn over a period of 2-4 weeks. A colony is in the right condition for getting combs drawn by this procedure when it has about 8 frames of brood in the deep box and a similar amount in the shallow box (see diagram Stage 1). It should be wall-to-wall packed with bees and in the stage of development when the beekeeper would be considering adding the first honey super.

Converting the brood nest into a tall, narrow configuration (Stage 2) makes it more heat efficient and creates optimum conditions for comb drawing in the upper box. Restricting brood nest expansion in the lower box by the strategic positioning of 2 well-filled frames of stores (or dummy boards if the former are not available) further encourages the colony to concentrate comb drawing in the 'sweet area' either side of the brood in the upper box. The aim is to re-peatedly move frames of foundation into a position on either side of the brood nest in the upper box which is the only place where expansion can occur. As soon as foundation has been drawn in the upper box there are two possible outcomes for the individual frames:-

Getting Comb Drawn – Stages 4-5

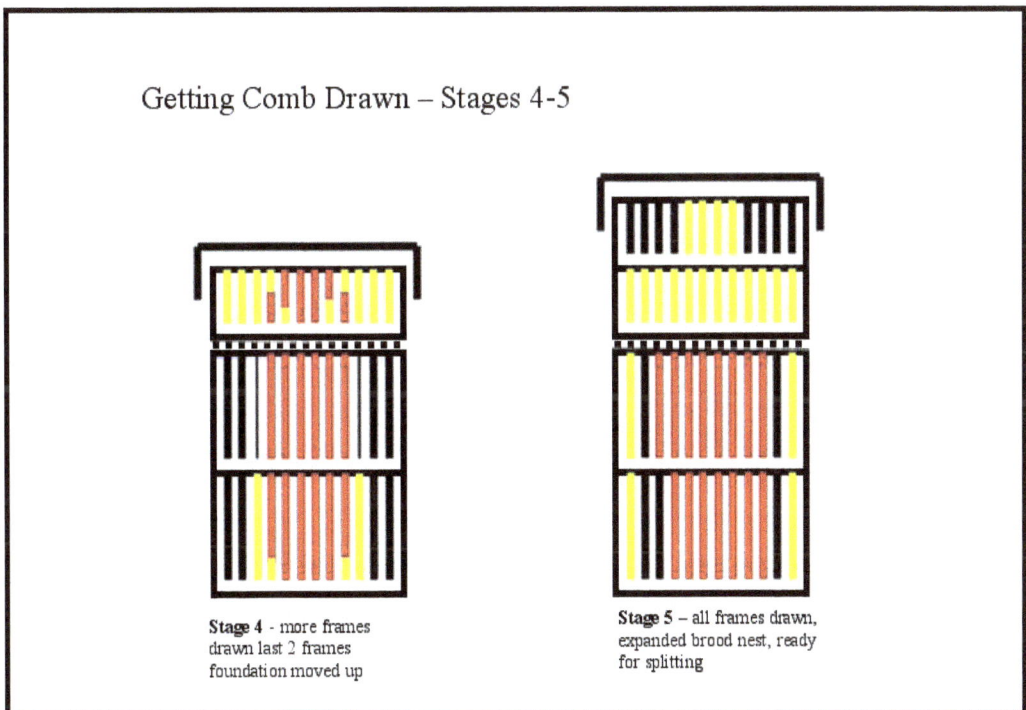

Stage 4 - more frames drawn last 2 frames foundation moved up

Stage 5 – all frames drawn, expanded brood nest, ready for splitting

Wally Shaw

1) Those that the queen has not yet laid in are initially transferred to the bottom box in exchange for frames of foundation. The newly drawn frames that are moved down should be placed outside the restricted brood nest area so that the queen is unable to lay in them. When all the foundation in the lower box has been exhausted, drawn frames (with no brood) should be placed to the outside of the upper box where they are separated from the brood nest by frames of foundation yet to be drawn.

2) The queen will inevitably lay in some frames as soon as they are drawn and these can either remain where they are in the top box or be moved down to widen the brood nest in the lower box. The aim should be to maintain a brood nest that is roughly egg-shaped (with its narrower end uppermost).

Frames that have been drawn on one side only can be turned through 180° (providing they have not been laid-in) to get the other side drawn more quickly. The process of moving frames of foundation, whilst at the same time allowing **controlled expansion** of the brood nest, is continued until drawing is complete as shown in Stage 5. The timing of inspections (5-7 days is suggested) and the progressive frame manipulations must be judged according to the rate of progress. On completion (Stage 5) not only will the colony have 12 additional drawn frames but the amount of brood will also have increased (from 8 to 14 frames in the example), ie. in an ideal condition for making increase.

The initial hive configuration shown in Stage 1 is a brood and half (with the half on top). The reason for moving the half-brood above a queen excluder (Stage 2) is so that the queen can not re-lay these frames. This forces the bees to concentrate on drawing the foundation to give her more space.

A similar method can also be applied to a hive on a single deep box. In one respect a single box is easier, because you do not have to find the queen, but it is also slightly less satisfactory. This is because the continuity of the brood into the half-brood contributes to the warmth in the upper box where the comb drawing is in progress.

The colour coding of the frames in the diagrams shows their **predominant characteristics**. In the real world, most frames will contain a mixture of brood, stores and empty cells in varying amounts and the detail of how they are deployed is a matter for the beekeeper's judgement (and it's not that critical).

Simple Methods of Making Increase

Unless there is a large nectar flow throughout most of the drawing period (which is fairly unlikely) feeding will be required. Bees will not draw comb unless there is a nectar flow or a simulated flow in the form of sugar syrup. Feeding in this situation should not result in contamination of the honey crop because there are no supers on the hive. As the brood in the half-brood emerges it is inevitable that some syrup will be stored in these frames. However, when the process of comb drawing is complete and it is time to split the hive, the half-brood will revert to its former role of holding brood and any transfer of sugar to the supers should be minimal.

19. Concluding Remarks

The process of making increase is both simple and extremely flexible and is within the capability of most beekeepers with as little as 1-2 year's experience. The above examples show just a few of the possibilities but it is the principles that really matter and once these are understood splits can be tailored to meet the beekeeper's needs.

Wally Shaw

www.ingramcontent.com/pod-product-compliance
Lightning Source LLC
LaVergne TN
LVHW061328060426
835511LV00012B/1921